The Hurricane Guide

by William Radcliff Birt

THE

HURRICANE GUIDE:

BEING

AN ATTEMPT TO CONNECT THE

ROTATORY GALE OR REVOLVING STORM

WITH

ATMOSPHERIC WAVES.

INCLUDING INSTRUCTIONS FOR OBSERVING THE PHᴇOMENA OF THE WAVES AND STORMS;

WITH

PRACTICAL DIRECTIONS FOR AVOIDING THE CENTRES OF THE LATTER.

BY

WILLIAM RADCLIFF BIRT.

LONDON: JOHN MURRAY, ALBEMARLE STREET. PUBLISHER TO THE ADMIRALTY. 1850.

PRINTED BY W. CLOWES AND SONS, STAMFORD STREET.

PREFACE.

In introducing the following pages to the notice of the Public, it is the Author's wish to exhibit in as clear a light as our present researches on the subjects treated of will allow, the connexion between one of the most terrific phenomena with which our globe is visited, and a phenomenon which, although but little known, appears to be intimately connected with revolving storms. How far he has succeeded, either in this particular object or in endeavouring to render the essential phenomena of storms familiar to the seaman, is left for the Public to determine. Should any advantage be found to result from the study of the Atmospheric Waves, as explained and recommended in this little work, or the seaman be induced by its perusal to attend more closely to the observations of those instruments that are calculated to warn him of his danger, an object will be attained strikingly illustrative of the Baconian aphorism, "Knowledge is Power."

Bethnal Green, April 19, 1849.

CONTENTS.

THE

HURRICANE GUIDE.

CHAPTER I.

PHENOMENA OF REVOLVING STORMS.

It is the object of the following pages to exhibit, so far as observation may enable us, and in as brief a manner as possible, the connexion, if any, that exists between those terrific meteorological phenomena known as "revolving storms," and those more extensive and occult but not less important phenomena, "atmospheric waves."

To the great body of our seamen, whether in her Majesty's or the mercantile service, the subject can present none other than the most interesting features. The laws that govern the transmission of large bodies of air from one part of the oceanic surface to another, either in a state of rapid rotation or presenting a more or less rectilineal direction, must at all times form an important matter of inquiry, and bear very materially on the successful prosecution of the occupation of the voyager.

In order to place the subjects above alluded to in such a point of view that the connexion between them may be readily seen, it will be important to notice the principal phenomena presented by each. Without going over the ground so well occupied by those able writers on the subject of storms-- Redfield, Reid, Piddington, and Thom--it will be quite sufficient for our present purpose simply to notice the essential phenomena of revolving storms as manifested by the barometer and vane. The usual indications of a storm in connexion with these instruments are the falling of the barometer and the freshening of the wind, and it is generally considered that a rapid fall of the mercury in the hurricane regions invariably precedes the setting in of a storm.

There are three classes of phenomena that present themselves to an observer, according as he is situated on the line or axis of translation, or in either the right or left hand semicircle of the storm. These will be rendered very apparent by a little attention to the annexed engraving, fig. 1.

In this figure the arrow-head is supposed to be directed true north, and the hurricane--as is the case in the American storms north of the 30th parallel--to be moving towards the N.E. on the line N.E.--S.W. If the ship take the hurricane with the wind S.E.,--the letters within the two larger circles indicating the direction of the wind in the storm according to the rotation as shown by the circle of arrow-heads, and which it is to be particularly noted is in the northern hemisphere contrary to the direction in which the hands of a watch move: in the southern hemisphere the rotation is reversed--the only phenomena presented by the storm are as follows:--The wind continues to blow from the S.E., increasing considerably in force with the barometer falling to a very great extent until the centre of the storm reaches the ship, when the fury of the winds is hushed, and a lull or calm takes place, generally for about half an hour, after which the wind springs up mostly with increased violence, but from the opposite quarter N.W., the barometer begins to rise, and as the storm passes off, the force of the wind abates.

The point to which we wish particularly to direct attention in connexion with this exposition of the phenomena attending the transmission of a storm is this:--If the observer so place himself at the commencement that the wind passes from his left hand towards his right, his face will be directed towards the centre of the storm; and the wind undergoing no change in direction, but only in force, will acquaint him with this important fact that the centre is not only gradually but surely approaching him: in other words, in the case before us, when he finds the wind from the S.E., and he places himself with his face to the S.W. he is looking towards the centre, and the wind rushes past him from his left to his right hand. Now the connexion of the barometer with this phase of the storm is falling with the wind from left to right, the observer facing the centre while the first half is transiting.[1] During the latter half these conditions are reversed, the observer still keeping his position, his face directed to the S.W., the barometer rises with a N.W. wind, which rushes past him from his right to his left hand with a decreasing force. We have therefore a rising barometer with the wind from right to left during the latter half of the storm, the observer having his back to the centre.

The above general enunciations of the barometric and anemonal phenomena of a rotating storm hold good with regard to the northern hemisphere, whatever may be the direction in which the hurricanes advance. This may be placed in a clearer light, as well as the remaining classes of phenomena shown, by consulting the following tables, constructed for the basin of the Northern Atlantic, and comparing them with fig. 1. In this basin, with lower latitudes than 25? the usual paths of the hurricanes are towards the north-west, in higher latitudes than 30?towards the north-east. The tables exhibit the veering of the wind with the movements of the barometer, according as the ship is situated in the right or left hand semicircle of the hurricane. It must here be understood that the right and left hand semicircles are determined by the observer so placing himself that his face is directed towards the quarter to which the hurricane is advancing.

LOWER LATITUDES.

NORTHERN HEMISPHERE.

Axis line, wind N.E., barometer falling, first half of storm. Axis line, wind S.W., barometer rising, last half of storm.

RIGHT-HAND SEMICIRCLE.

Wind E.N.E., E., E.S.E., S.E., barometer falling, storm increasing. Wind S.S.W., S., S.S.E., S.E., barometer rising, storm passing off.

LEFT-HAND SEMICIRCLE.

Wind N.N.E., N., N.N.W., N.W., barometer falling, storm increasing. Wind W.S.W., W., W.N.W., N.W., barometer rising, storm passing off.

HIGHER LATITUDES.

NORTHERN HEMISPHERE.[2]

Axis line, wind S.E., barometer falling, first half of storm. Axis line, wind N.W., barometer rising, last half of storm.

RIGHT-HAND SEMICIRCLE.

Wind S.S.E., S., S.S.W., S.W., barometer falling, storm increasing. Wind W.N.W., W., W.S.W., S.W., barometer rising, storm passing off.

LEFT-HAND SEMICIRCLE.

Wind E.S.E., E., E.N.E., N.E., barometer falling, storm increasing. Wind N.N.W., N., N.N.E., N.E., barometer rising, storm passing off.

N.B. The directions of the hurricane winds are so arranged as to show the points of commencement and termination. Thus in the lower latitudes a storm commencing at E.N.E. passes off at S.S.W. after the wind has veered E., E.S.E., S.E., S.S.E., and S., being in the order of the letters in the upper line and contrary to their order in the lower. One commencing at E.S.E. passes off at S.S.E. right-hand semicircle. In the higher latitudes a ship taking the storm at E.N.E. will be in the left-hand semicircle, and the hurricane will pass off at N.N.E. These changes are rendered very apparent by moving the hurricane circle in the direction in which the storm is expected to proceed.

Fig. 2 represents the whirl and hurricane winds in the south.

CHAPTER II.

PHENOMENA OF ATMOSPHERIC WAVES.

Professor Dove of Berlin has suggested that in the temperate zones the compensating currents of the atmosphere necessary to preserve its equilibrium may be arranged as parallel currents on the surface, and not superposed as in or near the torrid zone. His views may be thus enunciated:-- That in the parallels of central Europe the N.E. current flowing towards the equator to feed the ascending column of heated air is not compensated by a current in the upper regions of the atmosphere flowing from the S.W. as in the border of the torrid zone, but there are also S.W. currents on each side the N.E., which to the various countries over which they pass appear as surface-winds, the winds in fact being disposed in alternate beds or layers, S.W., N.E., as in fig. 3.

The Professor also suggests that these parallel and oppositely directed winds are shifting, i. e. they gradually change their position with a lateral motion in the direction of the large arrow cutting them transversely.

In the course of the author's researches on atmospheric waves he had an opportunity of testing the correctness of Professor Dove's suggestion, and in addition ascertained that there existed another set of oppositely directed winds at right angles to those supposed to exist by the Professor. These currents were N.W. and S.E. with a lateral motion towards the N.E. He also carefully discussed the barometric phenomena with relation to both these sets of currents, and arrived at the following conclusions. The details will be found in the author's third report, presented to the British Association for the Advancement of Science (Reports, 1846, pp. 132 to 162). During the period under examination the author found the barometer generally to rise with N.E. and N.W. winds, and fall with S.W. and S.E. winds, and that the phenomena might be thus illustrated:--Let the strata a a a' a', b' b' b b, fig. 3, represent two parallel aerial currents or winds, a a a' a' from S.W. or S.E., and b' b' b b from N.E. or N.W. and conceive them both to advance from the N.W. in the first instance and from the S.W. in the second, in the direction of the large arrow. Now conceive the barometer to commence rising just as the edge b b passes any line of country, and to continue rising until the edge b' b' arrives at that line, when the maximum is attained. It will be remarked that this rise is coincident with a N.E. or N.W. wind. The wind now changes and the barometer begins to fall, and continues falling until the edge a a coincides with the line of country on which b b first impinged. During this process we have all the phenomena exhibited by an atmospheric wave: when the edge b b passes a line of country the barometer is at a minimum, and this minimum has been termed the anterior trough. During the period the stratum b' b' b b transits, the barometer rises, and this rise has been called the anterior slope. When the conterminous edges of the strata a' a' b' b' pass, a barometric maximum extends along the line of country formerly occupied by the anterior trough, and this maximum has been designated the crest. During the transit of the stratum a' a' a a the barometer falls, and this fall has been characterised as the posterior slope; and when the edge a a occupies the place of b b, the descent of the mercurial column is completed, another minimum extends in the direction of the former, and this minimum has been termed the posterior trough.

It will be readily seen that the lateral passage of the N.W. and S.E. currents towards the N.E. presents precisely the same barometric and anemonal phenomena as the rotatory storms when moving in the same direction. If the observer, when the barometer is at a maximum with a N.W. wind, place himself in the same position with regard to the laterally advancing current as he did with regard to the advancing storm, i. e. with his face towards the quarter from which it is advancing--S.W., he will find that with a falling barometer and S.E. wind the current passes him from the left to the right hand; but if at a barometric minimum he place himself in the same position with his face directed to the quarter from which the N.W. current is advancing laterally, also S.W., he will find that with a rising barometer and N.W. wind the current passes him from right to left. Now the two classes of phenomena are identical, and it would not be difficult to show that, had we an instance of a rotatory storm in the northern hemisphere moving from N.W. to S.E., it would present precisely the same phenomena as to the direction of currents passing from left to right and from right to left with falling and rising barometers, increase and decrease in the force of the wind, &c., as the oppositely directed aerial currents do which pass over western central Europe.

In the absence of direct evidence of the production of a revolving storm from the crossing of two large waves, as suggested by Sir John Herschel, although it is not difficult to obtain such evidence, especially from the surface of the ocean, the identity of the two classes of phenomena exhibited by the storms and waves as above explained amounts to a strong presumption that there is a close connexion between them, and that a more minute investigation of the phenomena of atmospheric waves is greatly calculated to throw considerable light on the laws that govern the storm paths in both hemispheres. The localities in which these atmospheric movements, the waves, have been hitherto studied, have been confined to the northern and central parts of Europe--the west of Ireland, Alten in the north of Europe,

Lougan near the Sea of Azov, and Geneva, being the angular points of the included area. It will be remarked that the greatest portion of this area is inland, but there is one important feature which the study of the barometer has brought to light, and which is by no means devoid of significance, viz. that the oscillations are much greater in the neighbourhood of water, and this appears to indicate that the junction lines of land and water form by far the most important portions of the globe in which to study both the phenomena of storms and waves. It is also very desirable that our knowledge of these phenomena should, with immediate reference to the surface of the ocean, be increased, and in this respect captains and masters of vessels may render essential service by observing and recording the state of the barometer, and direction and force of the wind, several times in the course of the day and night;[3] and when it is considered that the immediate object in view is one in which the mariner is personally interested, and one in which, it may be, his own safety is concerned, it is hoped that the keeping of a meteorological register having especial reference to the indications of the barometer, and force and direction of the wind, will not be felt as irksome, but rather will be found an interesting occupation, the instruments standing in the place of faithful monitors, directing when and where to avoid danger, and the record furnishing important data whereby the knowledge of general laws may be arrived at, having an essential bearing on the interests of the service at large.

CHAPTER III.

OBSERVATIONS.

In sketching out a system of observation having especial reference to atmospheric waves and rotatory storms, regard has been had--first, to the instruments that should be used, the observations to be made with them, the corrections to be applied to such observations, and the form of registry most suitable for recording the results: second, to the times of observation: third, to the more important localities that should be submitted to additional observation: fourth, to peculiar phenomena requiring extraordinary observations for their elucidation: and fifth, to particular seasons, when the instruments should be watched with more than ordinary care.

The more important objects of observation having especial reference to atmospheric waves are those points which have been termed crests and troughs. These are simply the highest and lowest readings of the barometer, usually designated maxima and minima, and should for the object in view receive particular attention. Whenever there is reason to believe that the barometer is approaching either a maximum or minimum, additional observations should be resorted to, so as to secure as nearly as possible the precise time as reckoned at the ship, with her position, of its occurrence, as well as the altitude of the mercurial column at that time and place. By means of such observations as these on board several ships scattered over the surfaces of our great oceans, much valuable information may be accumulated of a character capable of throwing considerable light on the direction in which the lines of barometric maxima and minima stretch, and also a tolerably accurate notion may be formed of their progress, both as regards direction and rate. In immediate connexion with such observations particular attention should be paid to the direction of the wind according to the season.

SECTION I.--INSTRUMENTS.

Description and Position of Instruments.--The principal instrument requisite in these observations is the barometer, which should be of the marine construction, and as nearly alike as possible to those furnished to the Antarctic expedition which sailed under the command of Sir James Clark Ross. These instruments were similar to the ordinary portable barometers, and differed from them only in the mode of their suspension and the necessary contraction of the tubes to prevent oscillation from the motion of the ship. The barometer on shipboard should be suspended on a gimbal frame, which ought not to swing too freely, but rather so as to deaden oscillations by some degree of friction. To the upper portion of the tube in this construction of instrument light is alike accessible either in front or behind, and the vernier is furnished with a back and front edge, both being in precisely the same plane, nearly embracing the tube, and sliding up and down it by the motion of rackwork; by the graduation of the scale and vernier the altitude of the mercury can be read off to ?02 inch.

When the barometer is placed in the ship, its position should be as near midships as possible, out of the reach of sunshine, but in a good light for reading, and in a situation in which it will be but little liable to sudden gusts of wind and changes of temperature. Great care should be taken to ascertain the exact height of its cistern above the water-line, and in order to facilitate night observations every possible arrangement should be made for placing behind it a light screened by white paper.

Observations.--The first thing to be done is the reading off and recording the temperature indicated by the thermometer that in this construction of instrument dips into the mercury in the cistern. Sir John Herschel has suggested that "the bulb of the thermometer should be so situated as to afford the best chance of its indicating the exact mean of the whole barometric column, that is to say, fifteen inches above the cistern enclosed within the case of the barometer, nearly in contact with its tube, and with a stem so long as to be read off at the upper level."

Previous to making an observation with the barometer the instrument should be slightly tapped to free the mercury from any adhesion to the glass; any violent oscillation should, however, be carefully avoided. The vernier should then be adjusted to the upper surface of the mercury in the tube; for this purpose its back and front edges should be made to coincide, that is, the eye should be placed in exactly the same plane which passes through the edges; they should then be brought carefully down until they form a tangent with the curve produced by the convex surface of the mercury and the light is just excluded from between them and the point of contact. It is desirable in making this adjustment that the eye should be assisted by a magnifying-glass. The reading of the scale should then be taken and entered in the column appropriated to it in the proper form. If the instrument have no tubular or double-edged index, the eye should be placed carefully at the level of the upper surface of the mercury and the index of the vernier brought gently down to the same level so as apparently just to touch the surface, great care being taken that the eye index and surface of the mercury are all in the same plane.

Each observation of the barometer should be accompanied by an observation of the direction of the wind, which should be noted in the usual manner in which it is observed at sea. In connexion with the direction the force of the wind should be recorded in accordance with the following scale, contrived by Admiral Sir Francis Beaufort:--

0. Calm 1. Light air or just sufficient to give steerage way. 2. Light breeze { or that in which a well- } 1 to 2 knots. 3. Gentle breeze { conditioned man of war, } 3 to 4 knots. 4. Moderate breeze { with all sail set, and } 5 to 6 knots. { clean full, would go in } { smooth water, from } 5. Fresh breeze } { Royals, &c. 6. Strong breeze } { Single-reefed top-sails } { and top-gallant } or that in which such a { sails. 7. Moderate gale } ship could just carry in { Double-reefed } chase full and by { topsails, jib, &c. 8. Fresh gale } { Triple-reefed } { topsails, &c. 9. Strong gale } { Close-reefed top-sails } { and courses.

10. Whole gale or that with which she could scarcely bear close-reefed main topsail and reefed foresail. 11. Storm or that which reduces her to storm staysails. 12. Hurricane or that which no canvas could withstand.

Corrections.--As soon after the observations have been made as circumstances will permit, the reading of the barometer should be corrected for the relation existing between the capacities of the tube and cistern (if its construction be such as to require that correction), and for the capillary action of the tube; and then reduced to the standard temperature of 32?Fahr., and to the sea-level, if on shipboard. For the first correction the neutral point should be marked upon each instrument. It is that particular height which, in its construction, has been actually measured from the surface of the mercury in the cistern, and indicated by the scale. In general the mercury will stand either above or below the neutral point; if above, a portion of the mercury must have left the cistern, and consequently must have lowered the surface in the cistern: in this case the altitude as measured by the scale will be too short--vice vers? if below. The relation of the capacities of the tube and cistern should be experimentally ascertained, and marked upon the instrument by the maker. Suppose the capacity to be 1/50, marked thus on the instrument, "Capacity 1/50:" this indicates that for every inch of variation of the mercury in the tube, that in the cistern will vary contrariwise 1/50th of an inch. When the mercury in the tube is above the neutral point, the difference between it and the neutral point is to be reduced in the proportion expressed by the "capacity" (in the case supposed, divided by 50), and the quotient added to the observed height; if below, subtracted from it. In barometers furnished with a fiducial point for adjusting the lower level, this correction is superfluous, and must not be applied.

The second correction required is for the capillary action of the tube, the effect of which is always to depress the mercury in the tube by a certain quantity inversely proportioned to the diameter of the tube. This quantity should be experimentally determined during the construction of the instrument, and its amount marked upon it by the maker, and is always to be added to the height of the mercurial column, previously corrected as before. For the convenience of those who may have barometers, the capillary action of which has not been determined, a table of corrections for tubes of different diameters is placed in the Appendix, Table I.

The next correction, and in some respects the most important of all, is that due to the temperature of the mercury in the barometer tube at the time of observation, and to the expansion of the scale. Table II. of the Appendix gives for every degree of the thermometer and every half-inch of the barometer, the proper quantity to be added or subtracted for the reduction of the observed height to the standard temperature of the mercury at 32?Fahr.

After these the index correction should be applied. This is the amount of difference between the particular instrument and the readings of the Royal Society's flint-glass barometer when properly corrected, and is generally known as the zero. It is impossible to pay too much attention to the determination of this point. For this purpose, when practicable, the instrument should be immediately compared with the Royal Society's standard, and the difference of the readings of both instruments, when corrected as above, carefully noted and preserved. Where, however, this is impracticable, the comparison should be effected by means either of some other standard previously so compared, or of an intermediate portable barometer, the zero point of which has been well determined. Suspend the portable barometer as near as convenient to the ship's barometer, and after at least an hour's quiet exposure, take as many readings of both instruments as may be necessary to reduce the probable error of the mean of the differences below 0.001 inch. Under these circumstances the mean difference of all the readings will be the relative zero or index error, whence, if that of the intermediate barometer be known, that of the other may be found. As such comparisons will always be made when the vessel is in port, sufficient time can be allowed for making the requisite number of observations: hourly readings would perhaps be best, and they would have the advantage of forming part of the system when in operation, and might be accordingly used as such.

It is not only desirable that the zero point of the barometer should be well determined in the first instance; it should also be carefully verified on every opportunity which presents itself; and in every instance, previous to sailing, it should be re-compared with the standard on shore by the intervention of a portable barometer, and no opportunity should be lost of comparing it on the voyage by means of such an intermediate instrument with the standard barometers at St. Helena, the Cape of Good Hope, Bombay, Madras, Paramatta, Van Diemen's Island, and with any other instruments likely to be referred to as standards, or employed in research elsewhere. Any vessel having a portable barometer on board, the zero of which has been well determined, would do well, on touching at any of the ports above named, to take comparative readings with the standards at those ports, and record the differences between the standard, the portable, and the ship barometers. By such means the zero of one standard may be transported over the whole world, and those of others compared with it ascertained. To do so, however, with perfect effect, will require that the utmost care should be taken of the portable barometer; it should be guarded as much as possible from all accident, and should be kept safely in the "portable" state when not immediately used for comparison. To transport a well-authenticated zero from place to place is by no means a point of trifling importance. Neither should it be executed hurriedly nor negligently. Some of the greatest questions in meteorology depend on its due execution, and the objects for which these instructions have been prepared will be greatly advanced by the zero points of all barometers being referred to one common standard. Upon the arrival of the vessel in England, at the termination of the voyage, the ship's barometer should be again compared with the same standard with which it was compared previous to sailing; and should any difference be found, it should be most carefully recorded.

The correction for the height of the cistern above or below the water-line is additive in the former case, subtractive in the latter. Its amount may be taken, nearly enough, by allowing 0?01 in. of the barometer for each foot of difference of level.

An example of the application of these several corrections is subjoined:--

```
| Attached Therm. 54 胘 3. |Data for the correction of | | | the Instrument.
|    +---------------------------------+--------------------------------+    |Barometer
reading. 29?09 |Neutral point 30?23 | |Corr. for capacity - ?17 |Capacity
1/42 | | |Capillary action + ?32 | +-------------------------------------| | | 29?92
|Zero to Royal Society + ?36 | |Corr. for capillarity + ?32 |Corr. for altitude
above | | | water-line + ?04 | +-------------------------------------| | | 29?24 | |
|Corr. for temperature - ?68 | | +-------------------------------------| | | 29?56 | |
|Corr. for zero and water-line + ?40 | | +-------------------------------------| |
|Aggregate = pressure at | | | sea-level 29?96 | | +-----------------------------------
----+--------------------------------+
```

It would greatly facilitate the comparison of the barometric observations by projecting them in curves when all the proper corrections have been applied.

This may be accomplished by a much smaller expenditure of time than may at first be supposed. A paper of engraved squares on which the observations of twelve days may be laid down on double the natural scale, would be very suitable for the purpose.[4] The projection of each day's observations would occupy but a short time; and should circumstances on any occasion prevent the execution of it, when the ship was becalmed or leisure otherwise afforded, it would form an interesting and useful occupation, and serve to beguile some of the tedium often experienced at such intervals.

Registers.--For the particular object in view the register need not be very extensive. One kept in the annexed form will be amply sufficient. It should, however, be borne in mind that none but uncorrected observations should find admission; in point of fact it should be strictly a register of phenomena as observed, and on no account whatever should any entry be made from recollection, or any attempt made to fill up a blank by the apparent course of the numbers before and after. The headings of the columns will, it is hoped, be sufficiently explicit. It is desirable in practice that the column for remarks should embrace an entire page opposite the other entries, in order that occasional observations, as well as several other circumstances continually coming under review in the course of keeping a journal, may find entry.

METEOROLOGICAL REGISTER kept on board _____ during her voyage from ____ to ____ by ____.

```
+---------+----+------+-------+------+-------------------+--------+----------+ | | | | | |
Wind. | | | | | | | | Att. |-----------+------| | | | Date. |Lat.| Long.| Barom.|
Ther.| Direction.|Force.| Remarks| Observer.| |---------|----|------|-------|------
|-----------|------|--------|----------| | | |h. m.| | | | | | | | | | | | | | | | | | | | | |
| | | | | | | | | | | | | | | | | | | | | | | | | | | | | | | | | | | | | | | | | | | | +---+---
--+----+------+-------+------+-----------+------+--------+----------+
```

The only difference between the above form and one for the reception of corrected readings will be the dispensing with the column for the attached thermometer, and placing under the word Barom. "corrected."

II.--TIMES OF OBSERVATION.

There can be no question that the greatest amount of information, the accuracy of the data supplied, and in fact every meteorological element necessary to increase our knowledge of atmospheric waves, may be best obtained by an uninterrupted series of hourly observations made on board vessels from their leaving England until their safe arrival again at the close of their respective voyages; but from a variety of circumstances--the nature of the service in which the vessels may be employed, particular states of the weather, &c.--such a course of unremitting labour cannot be expected; it is therefore necessary to fix on some stated hours at which the instruments before particularized should be regularly observed throughout the voyage, and their indications faithfully recorded. The hours of 3 A.M., 9 A.M., 3 P.M., and 9 P.M., are now so generally known as meteorological hours, that nothing should justify a departure from them; and it is the more essential that these hours should be adopted in the present inquiry, because the series of observations made at intervals terminated by these hours can the more readily be used in connexion with those made contemporaneously on land, and will also serve to carry on investigations previously instituted, and which have received considerable illustration by means of observations at the regular meteorological hours; we therefore recommend their general adoption in all observations conducted at sea.

It is intended in the sequel to call attention to particular parts of the earth's surface where it is desirable that additional observations should be made, in order to furnish data of a more accurate character, and to mark more distinctly barometric changes than the four daily readings are capable of effecting. The best means of accomplishing this for the object in view appears to be the division of the interval of six hours into two equal portions, and to make the necessary observations eight times in the course of twenty-four hours. In the particular localities to which allusion has been made we recommend the following as the hours of observation:--

A.M. 3, 6, 9, noon. P.M. 3, 6, 9, midnight.

In other localities besides those hereafter to be mentioned, when opportunities serve, readings at these hours would greatly enhance the value of the four daily readings.

There are, however, portions of the surface of our planet, and probably also phenomena that occur in its atmosphere, which require still closer attention than the eight daily readings. One such portion would appear to exist off the western coast of Africa, and we recommend the adoption of hourly readings while sailing to the westward of this junction of aqueous and terrestrial surface; more attention will be directed to this point as we proceed. There are also phenomena the localities of which may be undetermined, and the times of their occurrence unknown, but so important a relation do they bear to the subject of our inquiries, that they demand the closest attention. They will be more particularly described under the head of accumulations of pressure preceding and succeeding storms, and minute directions given for the hourly observations of the necessary instruments. In the mean time we may here remark that hourly observations under the circumstances above alluded to are the more important when we consider that the barometer, the instrument employed in observing these moving atmospheric masses, is itself in motion. The ship may meet the accumulation of pressure and sail through it transversely; or she may sail along it, the course of the vessel being parallel to the line marking the highest pressure, the ridge or crest of the wave; or the ship may make any angle with this line: but whatever the circumstances may be under which she passes through or along with such an accumulation of pressure, it should ever be borne in mind that her position on the earth's surface is scarcely ever the same at any one observation as it was at the preceding, the barometer in the interval has changed its position as well as the line of maximum pressure, the rate of progress of which it is desirable to observe. It will, therefore, be at once apparent that in order to obtain the most accurate data on this head hourly observations are indispensable. To these readings should of course be appended the places of the ship from hour to hour, especially if she alter her course much.

There is another point to which we wish to call attention in immediate connexion with hourly readings--it is the observation of the instruments on the days fixed for that purpose: they were originally suggested by Sir John Herschel, whose directions should be strictly attended to: they are as follows:--

The days fixed upon for these observations are the 21st of March, the 21st of June, the 21st of September, and the 21st of December, being those, or immediately adjoining to those of the equinoxes and solstices, in which the solar influence is either stationary or in a state of most rapid variation. But should any one of those 21st days fall on a Sunday, then it will be understood that the observations are to be deferred till the next day, the 22nd. The series of observations on board each vessel should commence at 6 o'clock A.M. of the appointed days, and terminate at 6 A.M. of the days following, according to the usual reckoning of time adopted in the daily observations.

In addition to the twenty-five hourly readings at the solstices and equinoxes as above recommended, it would be desirable to continue the observations until a complete elevation and depression of the barometer had been observed at these seasons. This plan is adopted at the Royal Observatory, Greenwich, and would be attended with this advantage were it generally so-- the progress of the elevation and depression would be more readily traced and their velocities more accurately determined than from the four or eight daily readings.

III.--LOCALITIES FOR ADDITIONAL OBSERVATIONS.

In sketching out a system of barometric observation having especial reference to the acquisition of data from which the barometric character of certain large areas of the surface of the globe may be determined--inasmuch as such areas are distinguished from each other, on the one hand by consisting of extensive spaces of the oceanic surface unbroken, or scarcely broken, by land; on the other by the proximity of such oceanic surface to large masses of land, and these masses presenting two essentially different features, the one consisting of land particularly characterized as continental, the other as insular, regard has been accordingly had to such distribution of land and water.

As these instructions have especial reference to observations at sea, observations on land have not been alluded to; but in order that the data accumulated may possess that value which is essential for carrying on the inquiry in reference to atmospheric waves with success, provision is made to mark out more distinctly the barometric effects of the junction of large masses of land and water. It is well known that the oceanic surface, and even the smaller surfaces of inland seas, produce decided inflexions of the isothermal lines. They exercise an important influence on temperature. It has also been shown that the neighbourhood of water has a very considerable influence in increasing the oscillations of the mercurial column in the barometer, and in the great systems of European undulations it is well known that these oscillations increase especially towards the north-west. The converse of this, however, has not yet been subjected to observation; there has been no systematic co-operation of observers for the purpose of determining the barometric affections of large masses of water, such as the central portion of the basin of the northern Atlantic, the portion of oceanic surface between the Cape of Good Hope and Cape Horn, the Indian and Southern oceans, and the vast basin of the Pacific. Nor are we yet acquainted with the character of the oscillations, whether increasing or decreasing, as we recede from the central portions of the oceanic surfaces we have mentioned towards the land which forms their eastern, western, or northern boundaries. This influence of the junction line of land and water, so far as it is yet known, has been kept in view in framing these instructions, and, as it appears so prominently in Europe, it is hoped the additional observations between the four daily readings to which probably many observers may habitually restrict themselves, making on certain occasions and in particular localities a series of observations at intervals of three hours, will not be considered too frequent when the great importance of the problem to be solved is fully apprehended. It need scarcely be said that the value of these observations at three-hourly intervals will be greatly increased by the number of observers co-operating in them. Upon such an extensive system of co-operation a large space on the earth's surface, possessing peculiarities which distinguish it from others extremely unlike it in their general character, or assimilate it to such as possess with it many features in common, is marked out below for particular observation, occupying more than two-thirds of a zone in the northern hemisphere, having a breadth of 40? and including every possible variety of terrestrial and aqueous surface, from the burning sands of the great African desert, situated about the centre, to the narrow strip of land connecting the two Americas on the one side, and the chain of islands connecting China and Hindostan with Australia on the other. On each side of the African continent we have spaces of open sea between 30?and 40?west longitude north of the equator, and between 60?and 80?east longitude, in or to the south of the equator, admirably suited for contrasting the barometric affections, as manifested in these spaces of open water, with those occurring in situations where the influence of the terrestrial surface comes into more active operation.

The localities where three-hourly readings are chiefly desirable may be specified under the heads of Northern Atlantic, Southern Atlantic, Indian and Southern Oceans, and Pacific Ocean.

Northern Atlantic. Homeward-bound Voyages.--The discussion of observations made in the United Kingdom and the western border of central Europe, has indicated that off the north-west of Scotland a centre of great barometric disturbance exists. This centre of disturbance appears to be considerably removed from the usual tracks of vessels crossing the Atlantic; nevertheless some light may be thrown on the barometric phenomena resulting from this disturbance by observations during homeward-bound voyages, especially after the vessels have passed the meridian of 50?west longitude. Voyagers to or from Baffin and Hudson bays would do well during the whole of the voyage to read off the barometer every three hours, as their tracks would approach nearest the centre of disturbance in question. Before crossing the 50th meridian, the undulations arising from the distribution of land and water in the neighbourhood of these vast inland seas would receive considerable elucidation from the shorter intervals of observation, and after passing the 50th meridian the extent of undulation, as compared with that observed by the more southerly vessels, would be more distinctly marked by the three-hourly series. Surveying vessels stationed on the north-western coasts of Ireland and Scotland may contribute most important information on this head by a regular and, as far as circumstances will allow, an uninterrupted series either of six-hourly or three-hourly observations. The intervals of observation on board vessels stationed at the Western Isles, the Orkneys, and the Shetland Isles, ought not to be longer than three hours, principally on account of the great extent of oscillation observed in those localities. Vessels arriving from all parts of the world as they approach the United Kingdom should observe at shorter intervals than six hours. As a general instruction on this head the series of three-hourly observations may be commenced on board vessels from America and the Pacific by the way of Cape Horn on their passing the 20th meridian, such three-hourly observations to be continued until the arrival of the vessels in port. Ships by the way of the Cape of Good Hope should commence the three-hourly series either on leaving or passing the colony, in order that the phenomena of the tropical depression hereafter to be noticed may be well observed.

Northern Atlantic. Outward-bound Voyages.--Vessels sailing to the United States, Mexico, and the West Indies, should observe at three hours' interval upon passing the 60th meridian. Observations at this interval, on board vessels navigating the Gulf of Mexico and the Caribbean Sea, will be particularly valuable in determining the extent of oscillation as influenced by the masses of land and water in this portion of the torrid zone, as compared with the oscillation noticed off the western coast of Africa, hereafter to be referred to.

Southern Atlantic. Outward and homeward bound.--Without doubt the most interesting phenomenon, and one that lies at the root of the great atmospheric movements, especially those proceeding northwards in the northern hemisphere and southwards in the southern, is the equatorial depression first noticed by Von Humboldt and confirmed by many observers since. We shall find the general expression of this most important meteorological fact in the Report of the Committee of Physics and Meteorology, appointed by the Royal Society in 1840, as follows: "The barometer, at the level of the sea, does not indicate a mean atmospheric pressure of equal amount in all parts of the earth; but, on the contrary, the equatorial pressure is uniformly less in its mean amount than at and beyond the tropics." Vessels that are outward bound should, upon passing 40?north latitude, commence the series of three-hourly observations, with an especial reference to the equatorial depression. These three-hourly observations should be continued until the latitude of 40?south has been passed: the whole series will then include the minimum of the depression and the two maxima or apices forming its boundaries. (See Daniell's 'Meteorological Essays,' 3rd edition.) In passages across the equator, should the ships be delayed by calms, opportunities should be embraced for observing this depression with greater precision by means of hourly readings; and these readings will not only be valuable as respects the depression here spoken of, but will go far to indicate the character of any disturbance that may arise, and point out, as nearly as such observations will allow, the precise time when such disturbance produced its effects in the neighbourhood of the ships. In point of fact they will clearly illustrate the diversion of the tendency to rise, spoken of in the Report before alluded to, as resulting in ascending columns and sheets, between which wind flaws, capricious in their direction and intensity, and often amounting to sharp squalls, mark out the course of their feeders and the indraft of cooler air from a distance to supply their void. Hourly observations, with especial reference to this and the following head of inquiry, should also be made off the western coast of Africa during the homeward-bound voyage.

Immediately connected with this part of the outward-bound voyage, hourly observations, as often as circumstances will permit, while the ships are sailing from the Madeiras to the equator, will be extremely valuable in elucidating the origin of the great system of south-westerly atmospheric waves that traverse Europe, and in furnishing data for comparison with the amount of oscillation and other barometric phenomena in the Gulf of Mexico and the Caribbean Sea, a portion of the torrid zone essentially different in its configuration and in the relations of its area to land and water, as contra-distinguished to the northern portion of the African continent; and these hourly observations are the more desirable as the vessels may approach the land. They may be discontinued on passing the equator, and the three-hourly series resumed.

There are two points in the southern hemisphere, between 80?west longitude and 30?east longitude, that claim particular attention in a barometric point of view, viz., Cape Horn and the Cape of Good Hope; the latter is within the area marked out for the three-hourly observations, and too much attention cannot be paid to the indications of the barometer as vessels are approaching or leaving the Cape. The northern part of the South Atlantic Ocean has been termed the true Pacific Ocean of the world; and at St. Helena a gale was scarcely ever known; it is also said to be entirely free from actual storms (Col. Reid's 'Law of Storms,' 1st edition, p. 415). It may therefore be expected that the barometer will present in this locality but a small oscillation, and ships in sailing from St. Helena to the Cape will do well to ascertain, by means of the three-hourly observations, the increase of oscillation as they approach the Cape. The same thing will hold good with regard to Cape Horn: it appears from previous observation that a permanent barometric depression exists in this locality, most probably in some way connected with the immense depression noticed by Captain Sir James Clark Ross, towards the Antarctic Circle. The general character of the atmosphere off Cape Horn is also extremely different from its character at St. Helena. It would therefore be well for vessels sailing into the Pacific by Cape Horn, to continue the three-hourly observations until the 90th meridian is passed.

Before quitting the Atlantic Ocean it may be well to notice the marine stations mentioned in my Third Report on Atmospheric Waves,[5] as being particularly suitable for testing the views advanced in that report and for tracing a wave of the south-westerly system from the most western point of Africa to the extreme north of Europe. A series of hourly observations off the western coast of Africa has already been suggested. Vessels staying at Cape Verd Islands should not omit to make observations at three hours' interval during the whole of their stay, and when circumstances will allow, hourly readings. At the Canaries, Madeiras, and the Azores, similar observations should be made. Vessels touching at Cape Cantin, Tangier, Gibraltar, Cadiz, Lisbon, Oporto, Corunna, and Brest, should also make these observations while they are in the localities of these ports. At the Scilly Isles we have six-hourly observations, made under the superintendence of the Honourable the Corporation of the Trinity House. Ships in nearing these islands and making the observations already pointed out, will greatly assist in determining the increase of oscillation proceeding westward from the nodal point of the two great European systems. We have already mentioned the service surveying vessels employed on the coasts of Ireland and Scotland may render, and the remaining portion of the area marked out in the report may be occupied by vessels navigating the North Sea and the coast of Norway, as far as Hammerfest.

In connexion with these observations, having especial reference to the European system of south-westerly atmospheric waves, the Mediterranean presents a surface of considerable interest, both as regards these particular waves, and the influence its waters exert in modifying the two great systems of central Europe. The late Professor Daniell has shown from the Manheim observations, that small undulations, having their origin on the northern borders of the Mediterranean, have propagated themselves northward, and in this manner, but in a smaller degree, the waters of the Mediterranean have contributed to increase the oscillation as well as the larger surface of the northern Atlantic. In most of the localities of this great inland sea six-hourly observations may suffice for this immediate purpose; but in sailing from Lisbon through the Straits of Gibraltar, in the neighbourhood of Sicily and Italy, and in the Grecian Archipelago, we should recommend the three-hourly series, as marking more distinctly the effects resulting from the proximity of land; this remark has especial reference to the passage through the Straits of Gibraltar, where, if possible, hourly observations should be made.

The Indian and Southern Oceans. Outward and homeward bound.--On sailing from the Cape of Good Hope to the East Indies, China, or Australia, observations at intervals of three hours should be made until the 40th meridian east is passed (homeward-bound vessels should commence the three-hourly readings on arriving at this meridian). Upon leaving the 40th meridian the six-hourly observations may be resumed on board vessels bound for the Indies and China until they arrive at the equator, when the readings should again be made at intervals of three hours, and continued until the arrival of the vessels in port. With regard to vessels bound for Australia and New Zealand, the six-hourly readings may be continued from the 40th to the 100th meridian, and upon the vessels passing the latter, the three-hourly readings should be commenced and continued until the vessels arrive in port. Vessels navigating the Archipelago between China and New Zealand, should make observations every three hours, in order that the undulations arising from the configuration of the terrestrial and oceanic surfaces may be more distinctly marked and more advantageously compared with the Gulf of Mexico, the Caribbean Sea, and the northern portion of the African continent.

The Pacific Ocean.--As this ocean presents so vast an aqueous surface, generally speaking observations at intervals of six hours will be amply sufficient to ascertain its leading barometric phenomena. Vessels, however, on approaching the continents of North and South America, or sailing across the equator, should resort to the three-hourly readings, in order to ascertain more distinctly the effect of the neighbourhood of land on the oscillations of the barometer, as generally observed, over so immense a surface of water in the one case, and the phenomena of the equatorial depression in the other: the same remarks relative to the latter subject, which we offered under the head of South Atlantic, will equally apply in the present instance. The configuration of the western shores of North America renders it difficult to determine the precise boundary where the three-hourly series should commence; the 90th meridian is recommended for the boundary as regards South America, and from this a judgment may be formed as to where the three-hourly observations should commence in reference to North America.

In the previous sketch of the localities for the more important observations, it will be seen that within the tropics there are three which demand the greatest regard.

I. The Archipelago between the two Americas, more particularly comprised within the 40th and 120th meridians west longitude, and the equator and the 40th degree of north latitude. As a general principle we should say that vessels within this area should observe the barometer every three hours. Its eastern portion includes the lower branches of the storm paths, and on this account is peculiarly interesting, especially in a barometric point of view.

II. The Northern portion of the African Continent, including the Sahara or Great Desert.--This vast radiating surface must exert considerable influence on the waters on each side northern Africa. Vessels sailing within the area comprised between 40?west and 70?east, and the equator and the 40th parallel, should also make observations at intervals of three hours.

III. The great Eastern Archipelago.--This presents a somewhat similar character to the western; like that, it is the region of terrific hurricanes, and it becomes a most interesting object to determine its barometric phenomena; the three-hourly system of observation may therefore be resorted to within an area comprised between the 70th and 140th meridians, and the equator and the 40th degree of north latitude.

The southern hemisphere also presents three important localities, the prolongations of the three tropical areas. It is unnecessary to enlarge upon these, as ample instructions have been already given. We may, however, remark, with regard to Australia, that three-hourly observations should be made within the area comprised between the 100th and 190th meridians east, and the equator and the 50th parallel south, and hourly ones in the immediate neighbourhood of all its coasts.

IV.--STORMS, HURRICANES, AND TYPHOONS.

The solution of the question--How far and in what manner are storms connected with atmospheric waves?--must be extremely interesting to every one engaged in either the naval or merchant service. As we have in the former chapters directed attention to their connexion, our great object here will be to endeavour to mark out such a line of observation as appears most capable of throwing light, not only on the most important desiderata as connected with storms, but also their connexion or non-connexion with atmospheric waves. We shall accordingly arrange this portion of the instructions under the following heads:--Desiderata; Localities; Margins; Preceding and Succeeding Accumulations of Pressure.

Desiderata.--The most important desiderata appertaining to the subject of storms, are certainly their origin and termination. Of these initial and terminal points in the course of great storms we absolutely know nothing, unless the white appearance of a round form observed by Mr. Seymour on board the Judith and Esther, in lat. 17?19' north and long. 52?10' west (see Col. Reid's 'Law of Storms,' 1st edit. p. 65), may be regarded as the commencement of the Antigua hurricane of August 2, 1837. This vessel was the most eastern of those from which observations had been obtained; and it is the absence of contemporaneous observations to the eastward of the 50th meridian that leaves the question as to the origin of the West Indian revolving storms unsolved. Not one of Mr. Redfield's storm routes extends eastward of the 50th meridian; this at once marks out, so far as storms are concerned, the entire space included between the 20th and 50th meridians, the equator and the 60th parallel, as a most suitable area for observations, under particular circumstances hereafter to be noticed, with especial reference either to the commencement or termination of storms, or the prolongation of Mr. Redfield's storm paths.

Localities.--The three principal localities of storms are as follows:--I. The western portion of the basin of the North Atlantic; II. The China Sea and Bay of Bengal; and III. The Indian Ocean, more particularly in the neighbourhood of Mauritius. The first two have already been marked out as areas for the three-hourly observations; to the latter, the remark as to extra observations under the head of Desiderata will apply.

Margins.--Mr. Redfield has shown that on some occasions storms have been preceded by an unusual pressure of the atmosphere; the barometer has stood remarkably high, and it has hence been inferred that there has existed around the gale an accumulation of air forming a margin; barometers placed under this margin indicating a much greater pressure than the mean of the respective localities. With regard to the West Indian and American hurricanes--any considerable increase of pressure, especially within the space marked out to the eastward of the 50th meridian, will demand immediate attention. Upon the barometer ranging very high within this space, three-hourly observations should be immediately resorted to; and if possible, hourly readings taken, and this is the more important the nearer the vessel may be to the 50th meridian. Each observation of the barometer should be accompanied by an observation of the wind--its direction should be most carefully noted, and the force estimated according to the scale in page 21, or by the anemometer. It would be as well at the time to project the barometric readings in a curve even of a rough character, that the extent of fall after the mercury had passed its maximum might be readily discernible by the eye. A paper ruled in squares, the vertical lines representing the commencement of hours, and the horizontal tenths of an inch, would be quite sufficient for this purpose. The force of the wind should be noted at, or as near to the time of the passage of the maximum as possible. During the fall of the mercury particular attention should be paid to the manner in which the wind changes, should any change be observed; and should the wind continue blowing steadily in onedirection, but gradually increasing in force, then such increments of force should be most carefully noted. During the fall of the barometer, should the changes of the wind and its increasing force indicate the neighbourhood of a revolving storm, (independent of the obvious reasons for avoiding the focus of the storm,) it would contribute as much to increase our knowledge of these dangerous vortices to keep as near as possible to their margins as to approach their centres. The recess from the centre towards the margin of the storm, will probably be rendered apparent by the rising of the mercury; and so far as the observations may be considered valuable for elucidating the connexion of atmospheric waves with rotatory storms (other motives being balanced), it might be desirable to keep the ship near the margin--provided she is not carried beyond the influence of the winds which characterize the latter half of the storm--until the barometer has nearly attained its usual elevation. By this means some notion might be formed of the general direction of the line of barometric pressure preceding or succeeding a storm.

Should a gale be observed commencing without its having been preceded by an unusual elevation of the mercurial column, and consequently no additional observation have been made; when the force of the wind is noted in the usual observations at or above 5, then the three-hourly series should be resorted to, and the same care taken in noting the direction, changes, and force of the wind as pointed out in the preceding paragraph.

The foregoing remarks relate especially to the central and western portions of the North Atlantic; they will however equally apply to the remaining localities of storms. Under any circumstances, and in any locality, a high barometer not less than a low one should demand particular attention, and if possible, hourly readings taken some time before and after the passage of the maximum: this will be referred to more particularly under the next head.

Preceding and Succeeding Accumulations of Pressure.--Mr. Redfield has shown in his Memoir of the Cuba Hurricane of October, 1844, that two associated storms were immediately preceded by a barometric wave, or accumulation of pressure, the barometer rising above the usual or annual mean. We have just referred to the importance of hourly observations on occasions of the readings being high as capable of illustrating the marginal phenomena of storms, and in connexion with these accumulations of pressure in advance of storms we would reiterate the suggestion. These strips of accumulated pressure are doubtless crests of atmospheric waves rolling forwards. In some cases a ship in its progress may cut them transversely in a direction at right angles to their length, in others very obliquely; but in all cases, whatever section may be given by the curve representing the observations, too much attention cannot be bestowed on the barometer, the wet and dry bulb thermometer, the direction and force of the wind, the state of the sky, and the appearance of the ocean during the ship's passage through such an accumulation of pressure. When the barometer attains its mean altitude, and is rapidly rising above it in any locality, then hourlyobservations of the instruments and phenomena above noticed should be commenced and continued until after the mercury had attained its highest point and had sunk again to its mean state. In such observations particular attention should be paid to the direction and force of the wind preceding the barometric maximum--and the same phenomena succeeding it, and particular notice should be taken of the time when, and amount of any change either in the direction or force of the wind. It is by such observations as these, carried on with great care and made at every accessible portion of the oceanic surface, that we may be able to ascertain the continuity of these atmospheric waves, to determine somewhat respecting their length, to show the character of their connexion with the rotatory storm, and to deduce the direction and rate of their progress.

V.--SEASONS FOR EXTRA OBSERVATIONS.

In reference to certain desiderata that have presented themselves in the course of my researches on this subject (see Report of the British Association for the Advancement of Science, 1846, p. 163), the phasesof the larger barometric undulations, and the types of the various seasons of the year, demand particular attention and call for extra observations at certain seasons: of these, three only have yet been ascertained--the type for the middle of November--the annual depression on or about the 28th of November--and the annual elevation on or about the 25th of December. The enunciation of the first is as under: "That during fourteen days in November, more or less equally disposed about the middle of the month, the oscillations of the barometer exhibit a remarkably symmetrical character, that is to say, the fall succeeding the transit of the maximum or the highest reading is to a great extent similar to the preceding rise. This rise and fall is not continuous or unbroken; in some cases it consists of five, in others of threedistinct elevations. The complete rise and fall has been termed the great symmetrical barometric wave of November. At its setting in the barometer is generally low, sometimes below twenty-nine inches. This depression is generally succeeded by two well-marked undulations, varying from one to two days in duration. The central undulation, which also forms the apex of the great wave, is of larger extent, occupying from three to five days; when this has passed, two smaller undulations corresponding to those at the commencement of the wave make their appearance, and at the close of the last the wave terminates." With but slight exceptions, the observations of eight successive years have confirmed the general correctness of this type. On two occasions the central apex has not been the highest, and these deviations, with some of a minor character, form the exceptions alluded to. This type only has reference to London and the south-eastern parts of England; proceeding westward, north-westward, and northward, the symmetrical character of this type is considerably departed from; each locality possessing its own type of the barometric movements during November. The desiderata in immediate connexion with the November movements, as observed in the southern and south-eastern parts of England, that present themselves, are--the determination of the types for November, especially its middle portion, as exhibited on the oceanic surface within an area comprised between the 30th and 60th parallels, and the 1st and 40th meridians west. Vessels sailing within this area may contribute greatly to the determination of these types by making observations at intervals of three hours from the 1st of November to the 7th or 8th of December. The entire period of the great symmetrical wave of November will most probably be embraced by such a series of observations, as well as the annual depression of the 28th. For the elevation of the 25th of December the three-hourly observations should be commenced on the 21st, and continued until the 3rd or 4th of the succeeding January.

With respect to the great wave of November, our knowledge of it would be much increased by such a series of observations as mentioned above, being made on board surveying and other vessels employed off Scotland and Ireland; vessels navigating the North Sea; vessels stationed off the coasts of France, Spain, Portugal, and the northern parts of Africa, and at all our stations in the Mediterranean. In this way the area of examination would be greatly enlarged, and the differences of the curves more fully elucidated; and this extended area of observation is the more desirable, as there is some reason to believe that the line of greatest symmetry revolves around a fixed point, most probably the nodal point of the great European systems.

It is highly probable that movements of a somewhat similar character, although presenting very different curves, exist in the southern hemisphere. The November wave is more or less associated with storms. It has been generally preceded by a high barometer and succeeded by a low one, and this low state of the barometer has been accompanied by stormy weather. We are therefore prepared to seek for similar phenomena in the southern hemisphere, in those localities which present similar states of weather, and at seasons when such weather predominates. We have already marked out the two capes in the Southern hemisphere for three-hourly observations: they must doubtless possess very peculiar barometric characters, stretching as they do into the vast area of the Southern Ocean. It is highly probable that the oscillations, especially at some seasons, are very considerable, and vessels visiting them at such seasons would do well to record with especial care the indications of the instruments already alluded to. At present we know but little of the barometric movements in the Southern hemisphere, and every addition to our knowledge in this respect will open the way to more important conclusions.

It has been observed in the south-east of England that the barometer has generally passed a maximum on or about the 3rd of every month, and this has been so frequently the case as to form the rule rather than the exception. The same fact during a more limited period has been observed at Toronto. With especial reference to this subject the three-hourly series of observations may be resorted to in all localities, but especially north of the 40th parallel in the northern hemisphere. They should be commenced at midnight immediately preceding the 1st and continued to midnight succeeding the 5th.

CHAPTER IV.

PRACTICAL DIRECTIONS FOR AVOIDING THE CENTRES OF ROTATING STORMS.

Figures 1 and 2, enlarged and printed on narrow rings of stiff cardboard, are employed for this purpose. The letters outside the thick circle are intended to distinguish the points of the compass, and in use should always coincide with those points on the chart. The letters within the thick circle indicate the direction of the wind in a hurricane, the whirl being shown by the arrows between the letters. In the northern hemisphere the direction of the whirl is always contrary to that in which the hands of a watch move, and in the southern coincident thereto. The graduation is intended to assist the mariner in ascertaining the bearing of the centre of a storm from his ship.

Use.

At any time when a severe gale or hurricane is expected, the seaman should at once find the position of his ship on the chart, and place upon it the graduated point which answers to the direction of the wind at the time, taking care that the needle is directed to the north, so that the exterior letters may point on the chart to the respective points of the compass: this is very essential. This simple process will at once acquaint the seaman with two important facts relative to the coming hurricane--his position in the storm, and the direction in which it is moving.

Examples.

A captain of a ship in latitude 35?24' N., longitude 64?12' W., bound to the United States, observes the barometer to stand unusually high, say 30?5 inches: shortly after the mercury begins to fall, at first slowly and steadily; as the glass falls the wind freshens, and is noticed to blow with increasing force from the S. so as to threaten a gale. The position of the ship on the chart is now to be found, and the graduated point under the letters E. S. is to be placed thereon, taking care to direct the needle to the north. From these two circumstances, the falling barometer and the wind blowing from the south with increasing force, the mariner is aware of this simple fact, that he is situated in the advancing portion of a body of air which is proceeding towards the N.E.; and if he turn his face to the N.E. he will find he is on the right of the axis line, or line cutting the advancing body transversely. The hurricane circle as it lies on the chart reveals to him another important fact, which is, that if he pursue his course he will sail towards the axis line of the hurricane, and may stand a chance of foundering in its centre. To avoid this he has one of two courses to adopt; either to lay-to on the starboard tack, according to Col. Reid's rules (see his 'Law of Storms,' 1st edit., pp. 425 to 428), the ship being in the right-hand semicircle of the hurricane, or so to alter his course as to keep without the influence of the storm. In the present case the adoption of the latter alternative would involve a reversal of his former course; nevertheless it is clear the more he bears to the S.E. the less he will experience the violence of the hurricane: should he heave his ship to, upon moving the hurricane circle from the ship's place on the chart towards the N.E., he will be able to judge of the changes of the wind he is likely to experience: thus it will first veer to S.S.W., the barometer still falling; then to S.W., the barometer at a minimum--this marks the position of the most violent portion of the storm he may be in, and by keeping the barometer as high as he can by bearing towards the S.E., the farther he will be from the centre--the barometer now begins to rise, the wind veering to W.S.W., and the hurricane finally passes off with the wind at W. It is to be particularly remarked that in this example the ship is in the most dangerous quadrant, as by scudding she would be driven in advance of the track of the storm's centre, which of course would be approaching her.

Assuming that the hurricane sets in at the ship's place with the wind at S.E., the proceeding will be altogether different. At first the wind is fair for the prosecution of the voyage, and it is desirable to take advantage of this fair wind to avoid as much as possible the track of the centre, which passes over the ship's place in this instance, and is always the most dangerous part of the storm. As the ship is able to make good distance from this track by bearing towards the N.W., provided she has plenty of sea-room, she will experience less of the violence of the hurricane; but as most of the Atlantic storms sweep over the shore, it will be desirable to lay-to at some point on the larboard tack, the ship being now in the left-hand semicircle. By moving the circle as before directed it will be seen that the veering of the wind is now E.S.E., E., E.N.E., N.E., the lowest barometer N.N.E., N., and N.N.W., the ship experiencing more or less of these changes as it is nearer to or farther from the axis line.

In latitudes lower than 20?N. the Atlantic hurricanes usually move towards the N.W. Taking the same positions of our ship with regard to the storms as in the two former examples, if the storm set in with the wind E. the proper proceeding is to bear away for the N.E., the most dangerous quadrant of the hurricane having overtaken the ship, the veering of the wind if she is lying-to will be E., E.S.E., S.E., with the lowest barometer S.S.E. and S. Should the storm set in at N.E., her position at the time will be some indication of the distance of the centre's track from the nearest land, and will greatly assist in determining the point at which the captain ought to lay-to after taking advantage of the N.E. wind, should he be able so to do, to bear away from the centre line, so as to avoid as much as possible the violence of the storm. From the proximity of the West Indian Islands to this locality of the storm-paths, the danger is proportionally increased.

The above examples have reference only to the lower and upper branches of the storm paths of the Northern Atlantic in the neighbourhood of the West Indies and the United States. In latitudes from about 25?to 32?these paths usually re-curve, and at some point will move towards the north. The veering of the wind will consequently be more or less complicated according as the ship may be nearer to or farther from the centre. The tables on page 11, combined with the first of those immediately following the next paragraph, will, it is hoped, prove advantageous in assisting the mariner as to the course to be adopted. As a general principle we should say it would be best to bear to the eastward, so as not only to avoid the greater fury of the storm, but to get into the S. and S.W. winds, which give the principal chances of making a westerly course.

We have in page 44 called attention to the fact that the storm paths traced by Mr. Redfield do not extend eastward of the 50th meridian. This by no means precludes the existence of severe storms and those of a rotatory character in the great basin of the Northern Atlantic, especially between the 40th and 50th parallels. A remarkable instance has come under the author's attention of the wind hauling apparentlycontrary to the usual theory: it may be that the storm route was in a direction not generally observed. We are at the present moment destitute of any information that at all indicates a reversion of the rotation in either hemisphere. The following tables constructed for the northern hemisphere, and for storm routes not yet ascertained, may probably be consulted with advantage on anomalous occasions.

HURRICANE MOVING FROM SOUTH TO NORTH.

Axis line, wind E., barometer falling, first half of storm. Axis line, wind W., barometer rising, last half of storm.

RIGHT-HAND SEMICIRCLE.

Wind E.S.E., S.E., S.S.E., S., barometer falling, first half of storm. Wind W.S.W., S.W., S.S.W., S., barometer rising, last half of storm.

LEFT-HAND SEMICIRCLE.

Wind E.N.E., N.E., N.N.E., N., barometer falling, first half of storm. Wind W.N.W., N.W., N.N.W., N., barometer rising, last half of storm.

HURRICANE MOVING FROM NORTH TO SOUTH.

Axis line, wind W., barometer falling, first half of storm. Axis line, wind E., barometer rising, last half of storm.

RIGHT-HAND SEMICIRCLE.

Wind W.N.W., N.W., N.N.W., N., barometer falling, first half of storm. Wind E.N.E., N.E., N.N.E., N., barometer rising, last half of storm.

LEFT-HAND SEMICIRCLE.

Wind W.S.W., S.W., S.S.W., S., barometer falling, first half of storm. Wind E.S.E., S.E., S.S.E., S,, barometer rising, last half of storm.

HURRICANE MOVING PROM WEST TO EAST.

Axis line, wind S., barometer falling, first half of storm. Axis line, wind N., barometer rising, last half of storm.

RIGHT-HAND SEMICIRCLE.

Wind S.S.W., S.W., W.S.W., W., barometer falling, first half of storm. Wind N.N.W., N.W., W.N.W., W., barometer rising, last half of storm.

LEFT-HAND SEMICIRCLE.

Wind S.S.E., S.E., E.S.E., E., barometer falling, first half of storm. Wind N.N.E., N.E., E.N.E., E., barometer rising, last half of storm.

HURRICANE MOVING FROM NORTH-WEST TO SOUTH-EAST.

Axis line, wind S.W., barometer falling, first half of storm. Axis line, wind N.E., barometer rising, last half of storm.

RIGHT-HAND SEMICIRCLE.

Wind W.S.W., W., W.N.W., N.W., barometer falling, first half of storm. Wind N.N.E., N., N.N.W., N.W., barometer rising, last half of storm.

LEFT-HAND SEMICIRCLE.

Wind S.S.W., S., S.S.E., S.E., barometer falling, first half of storm. Wind E.N.E., E., E.S.E., S.E., barometer rising, last half of storm.

###

www.ingramcontent.com/pod-product-compliance
Lightning Source LLC
Chambersburg PA
CBHW070924180526
45168CB00005B/2143